CORDOVA BRANCH LIBRARY

Solar Storms

The Rosen Publishing Grou
READING ROOM
Colle
New York

Published in 2003 by The Rosen Publishing Group, Inc.
29 East 21st Street, New York, NY 10010

Copyright © 2003 by The Rosen Publishing Group, Inc.

First Library Edition 2003

All rights reserved. No part of this book may be reproduced in any form without permission in writing from the publisher, except by a reviewer.

Book Design: Ron A. Churley

Photo Credits: Cover, p. 1 © PhotoDisc; pp. 4–5 © Great Lakes, Sunrise, USA/International Stock Photography; pp. 6–7 © GSO Images/The Image Bank; pp. 8–9 © Digital Vision; pp. 10–11, 11 (insets), 12–13, 15, 15 (top and bottom left insets), 16 (insets), 19 (center inset) © PhotoDisc; p. 15 (bottom right inset) © Finley-Holiday/FPG International; pp. 16–17, 19 (top inset) © Steven Nourse/Stone; p. 19 (bottom inset) © Getty's Digitized NASA Images/The Image Bank; p. 20 © AFP/Corbis; p. 21 © Bettmann/Corbis; pp. 22–23 © Wayne R. Bilenduke/Stone.

Library of Congress Cataloging-in-Publication Data

Stewart, Tobi Stanton.
Solar storms / Tobi Stanton Stewart.
p. cm. — (The Rosen Publishing Group's reading room collection)
Includes index.
Summary: Examines the characteristics and importance of the sun and how sunspots, solar wind, and solar storms affect the earth.
ISBN 0-8239-3709-7 (library binding)
1. Sun—Juvenile literature. 2. Solar wind—Juvenile literature. [1. Sun.] I. Title. II. Series.
QB521.5 .S84 2003
523.7'2—dc21

2001008067

Manufactured in the United States of America

For More Information

University of California/Exploratorium
http://www.exploratorium.edu/learning_studio/auroras/

Alaska Space Science Adventures (University of Alaska)
http://dac3.pfrr.alaska.edu/~ddr/ASGP/ELEMPART/INDEX.HTM

Contents

Our Sun

There would be no life on the Earth without the Sun. If the Sun did not exist, our world would be ice-covered, dark, and lifeless. We need the Sun for light, heat, and energy. Plants need the Sun to grow, and animals depend on plants for food.

The Sun is located at the center of our **solar system**. It is a huge star made up of hot gases. The Sun is about 93 million miles away from Earth. If you were able to drive to the Sun in a car going about sixty miles an hour, it would take you about 180 years to get there!

The Sun is about 4.5 billion years old, but you don't have to worry about the Sun burning out any time soon. Scientists believe the Sun will last for about another 5 billion years!

The Sun is the star that is closest to Earth.

Think of how warm the Sun makes you feel on a hot summer day. Now imagine how hot the Sun actually is. The Sun is so hot that solid matter can't exist there. It is made entirely of gases. The surface of the Sun is the part we can see from Earth. This surface is about 10,000 degrees **Fahrenheit**. That is about forty-seven times hotter than boiling water. However, that is not the hottest part of the Sun. The Sun's center, or core, is about 27 million degrees!

All that heat at the Sun's core makes 5 million tons of pure energy every second! After about 1 million years, that energy reaches the Sun's surface and explodes with great force.

The Sun is always changing. It burns and releases gases all the time. These changes affect our weather. By studying the Sun, we can learn more about these changes.

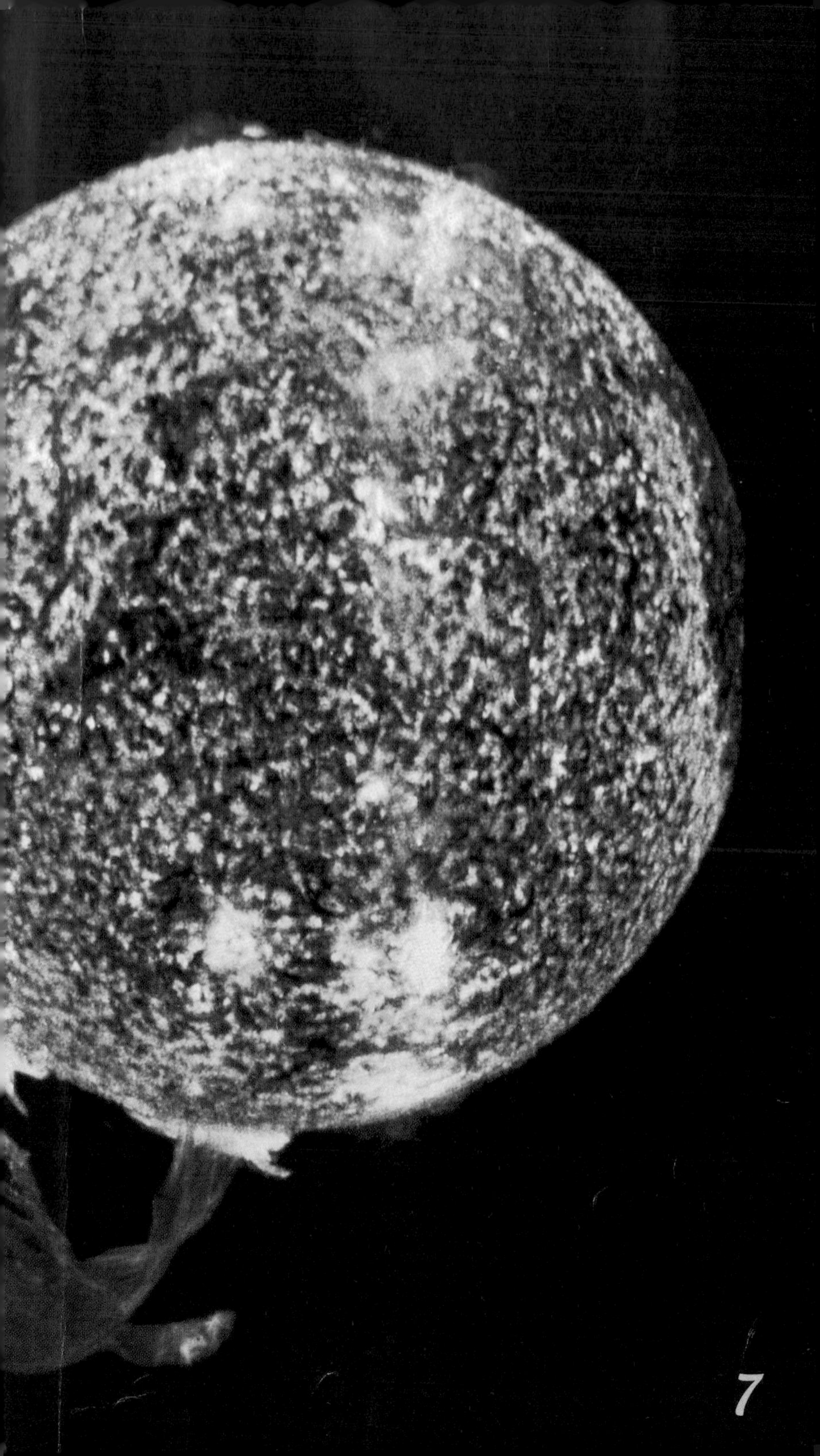

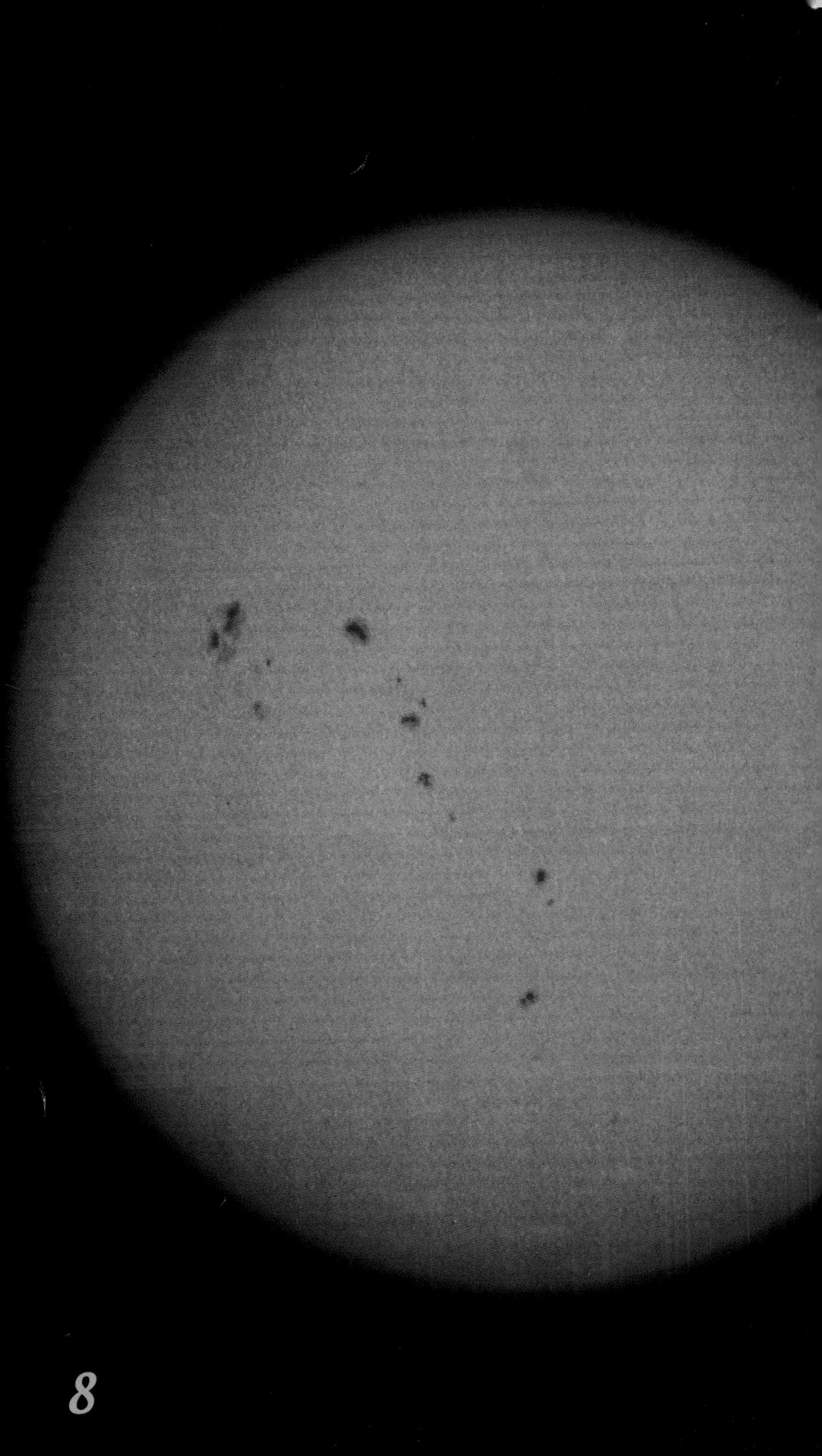

Solar Storms

Our space weather begins with the Sun. The Sun is always changing because its gases move around a lot. Sometimes this causes sunspots. Sunspots are places on the Sun's surface that are cooler than the areas around them. They appear as dark spots on the Sun's surface. Sunspots can measure up to 31,000 miles across! That is almost four times as wide as Earth!

Sunspots cause solar storms, which affect our space weather. When the Sun has more sunspots, solar storm activity increases. The number of sunspots on the Sun's surface is always changing. It takes about eleven years for he number of sunspots on the Sun's surface o increase from about five sunspots to about 00 sunspots. This eleven-year period is called he "sunspot cycle."

Most sunspots measure about 20,000 miles across. That is more than two times as wide as Earth!

During solar storms, gases explode from the Sun's surface with great force. These explosions are called **solar flares**. When the Sun has a lot of sunspots, it has even more solar flares than usual. A solar flare is about 20,000 degrees Fahrenheit. That's about twice as hot as the Sun's surface! Smaller solar flares may last for about ten minutes. The largest solar flares can last for about an hour.

Solar flares release energy in the form of light, heat, **radiation** (ray-dee-AY-shun), and very tiny, fast-moving **particles** called cosmic (KOZ-mihk) rays. Cosmic rays can affect our radio signals on Earth. An increase in radiation can harm people's skin.

Scientists think that since the formation of our solar system, the amount of energy the Sun releases has increased by about 40 percent!

The Solar Wind

The solar wind is made of gases that explode from the Sun's surface during solar storms. These gases contain electrically charged particles, which flow out into space towards all of the planets in our solar system, including Earth.

The space around Earth is called our **atmosphere** (AT-moh-sfeer). Earth's atmosphere is made up of gases and dust. There is also a **magnetic field** around Earth. We can't see this magnetic field, but it is there.

The center of Earth's core acts just like a magnet. If you've ever used magnets in science class, you know how a magnet can pull, push, and hold things around it. When the charged particles from the solar wind enter Earth's atmosphere, they cause changes in Earth's magnetic field.

Earth

When the solar wind reaches Earth's atmosphere, it is moving at a speed of up to 2 million miles per hour!

Solar flares can affect the solar wind. When solar flares explode from the Sun's surface, they send fast-moving particles into the solar wind. This makes the solar wind move faster and press even harder on Earth's atmosphere. This causes changes in Earth's magnetic field that can have powerful effects on our everyday lives.

For instance, changes in Earth's magnetic field can interfere with phone lines. Changes in Earth's magnetic field also increase the amount of power flowing through electrical lines. This can overload power lines and cause a **blackout**. Radio signals and **satellites** can also be affected.

During solar storms, even compass needles can be affected by changes in Earth's magnetic field!

radio tower
compass
satellite

Most auroras look like colorful clouds and streaks of light in the sky. They often look like they are moving or flickering.

Auroras

When the solar wind hits Earth's magnetic field, it can't get through it because the solar wind's charged particles are pushed away by Earth's charged particles. This is the same thing that happens when a magnet pushes away a similarly charged magnet. The solar wind's particles are forced to flow around Earth.

These particles interact with the gases in Earth's atmosphere to make different colors of light. This light is called an aurora (uh-ROAR-uh). The color of an aurora's light depends on which kind of gas mixes with the solar wind. Every gas glows with its own special color. Auroras are usually greenish in color, but they can also be red, purple, or blue. Auroras can only be seen at night when the sky is dark. They are the most easily seen examples of how solar storms affect Earth.

Auroras usually occur near Earth's North and South Poles. This is where Earth's magnetic field is strongest. The aurora seen around the North Pole is called aurora borealis (boor-ee-AL-iss). This aurora is also called the "northern lights." The aurora seen around the South Pole is called aurora australis (aw-STRAL-iss), or the "southern lights."

The best places to see auroras are near Earth's magnetic poles. These include places like Greenland, Siberia, and Alaska in the north. In the south, you'd have to travel to Antarctica to see an aurora. From late March until September, the North Pole has sunlight twenty-four hours a day, making auroras almost impossible to see.

During very active times in the sunspot cycle, big changes in Earth's magnetic field can make auroras shift away from Earth's poles toward the equator.

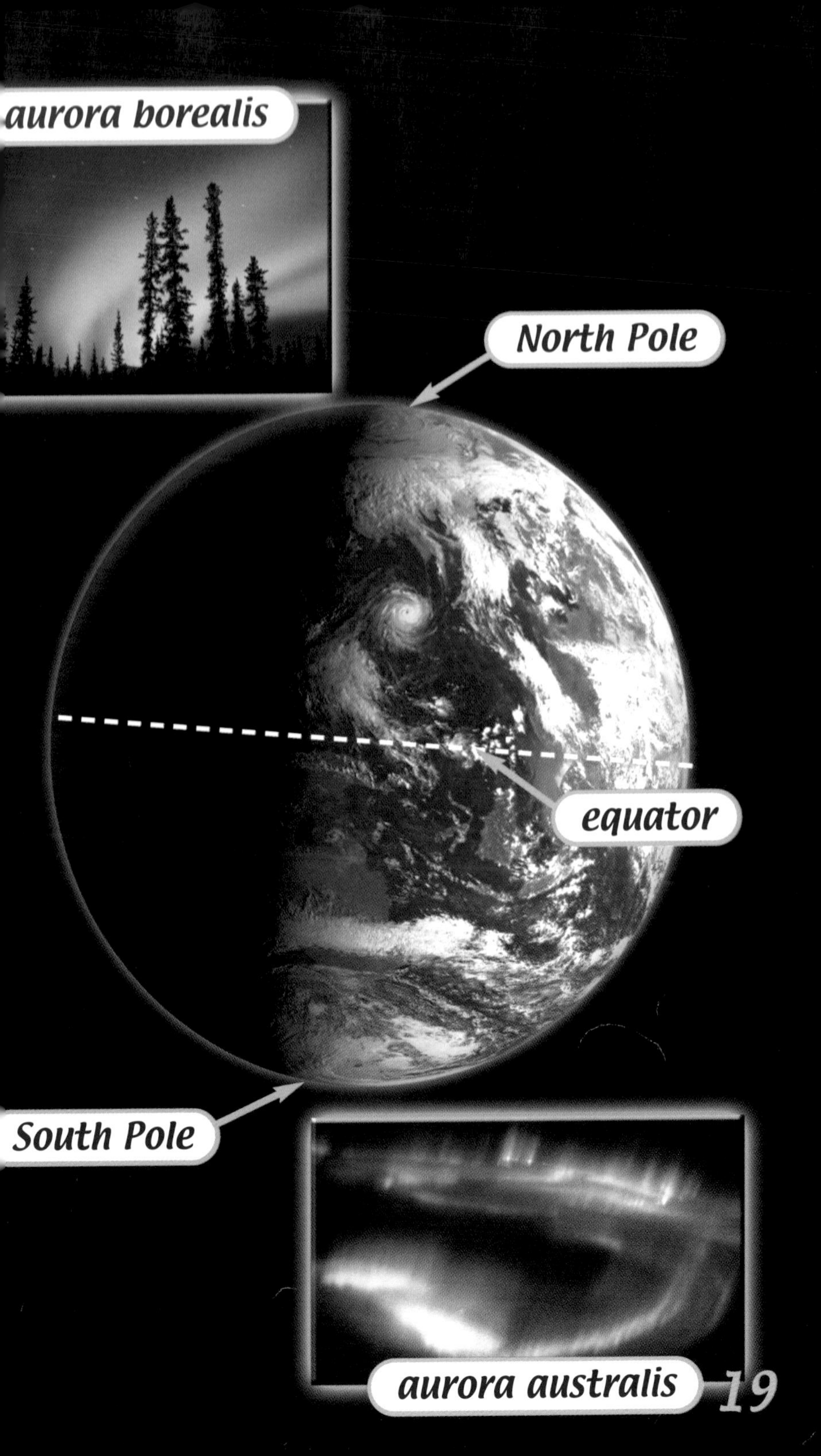
aurora borealis
North Pole
equator
South Pole
aurora australis

Most auroras happen anywhere from sixty to 600 miles above Earth. Some auroras can spread across the sky for thousands of miles. Today we understand that solar storms cause auroras, but throughout history, different cultures have had different beliefs about auroras.

Hundreds of years ago, people believed red auroras meant a war would soon occur. The reddish color stood for the blood that men would shed during battle.

Long ago, some Native Americans believed the northern lights were fires that were lit by spirits who lived in the sky. A thousand years ago in Europe, people believed the lights were actually the breath of men who had died in battles. Some people thought auroras were a sign of future events. Some people even believed that you shouldn't try to scare the northern lights by waving, whistling, or staring!

Why Do Solar Storms Happen?

In 1610, an Italian scientist named Galileo (gah-lih-LAY-oh) discovered sunspots with a new invention called the **telescope**. Since then we have learned many things about the Sun and solar storms. The strongest telescope ever built, called the Hubble Space Telescope, orbits Earth and shows us things about the Sun and space we could not know without it.

However, there are still many things we don't understand. We do not know why sunspots have eleven-year cycles. We do not know why solar flares happen. We also cannot tell when the next flare will occur, where it will occur, or how big it will be. Finding the answers to these questions will keep scientists busy for a very long time!

Glossary

atmosphere	The layer of air, gases, and dust that surrounds Earth.
blackout	A sudden loss of electrical power that makes lights go off and makes machines stop working.
Fahrenheit	A system used to measure how hot or cold something is.
magnetic field	The space around a magnet or electric current in which magnetic force happens.
particle	A very small piece of something.
radiation	Rays of energy from the Sun.
satellite	A machine that travels through space and is used to study objects in our solar system.
solar flare	An explosion of gas from the Sun's surface.
solar system	The system made up of our Sun, the nine planets, moons, and other space objects.
telescope	An instrument you look through that makes things that are far away look bigger and closer.

Index